Large Book of Laura Print Dot to Dot Therapy for Adults
From 150-636 Dots

By Laura's Dot to Dot Therapy

How To Use This Book

Hi! We're so glad you're a lover of puzzles and dot connecting- we are too!

Connecting the dots in this book is simple- just relax and follow the numbers in consecutive order, drawing a straight line between each one. Dot 1 will connect to dot 2 and so on and so forth until there are no more dots to connect. There's always another dot and you'll always find it. Connect every dot to discover the beautiful images they create.

In case you get lost or can't find a dot, never stress- there's an answer key at the back of the book that will show you exactly where each dot connects to the next. If you want to color your images, we encourage you to do so! Feel free to try all different colors and coloring mediums for your images!

If you find any errors or omissions in this book, email us at Laurasdottodot@gmail.com and please let us know! We want you to have the best dot to dot experience!

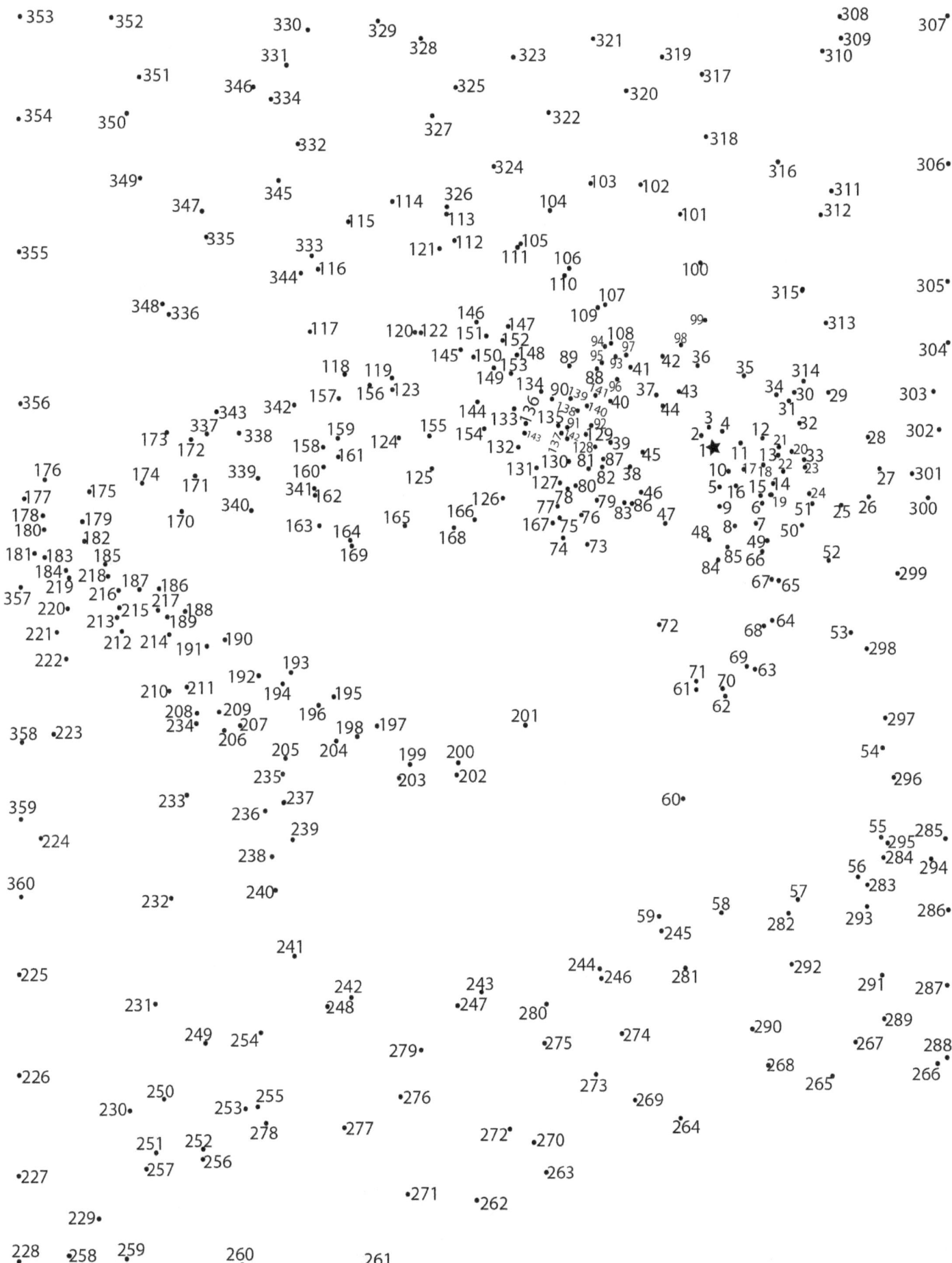

Page 2

Page 3

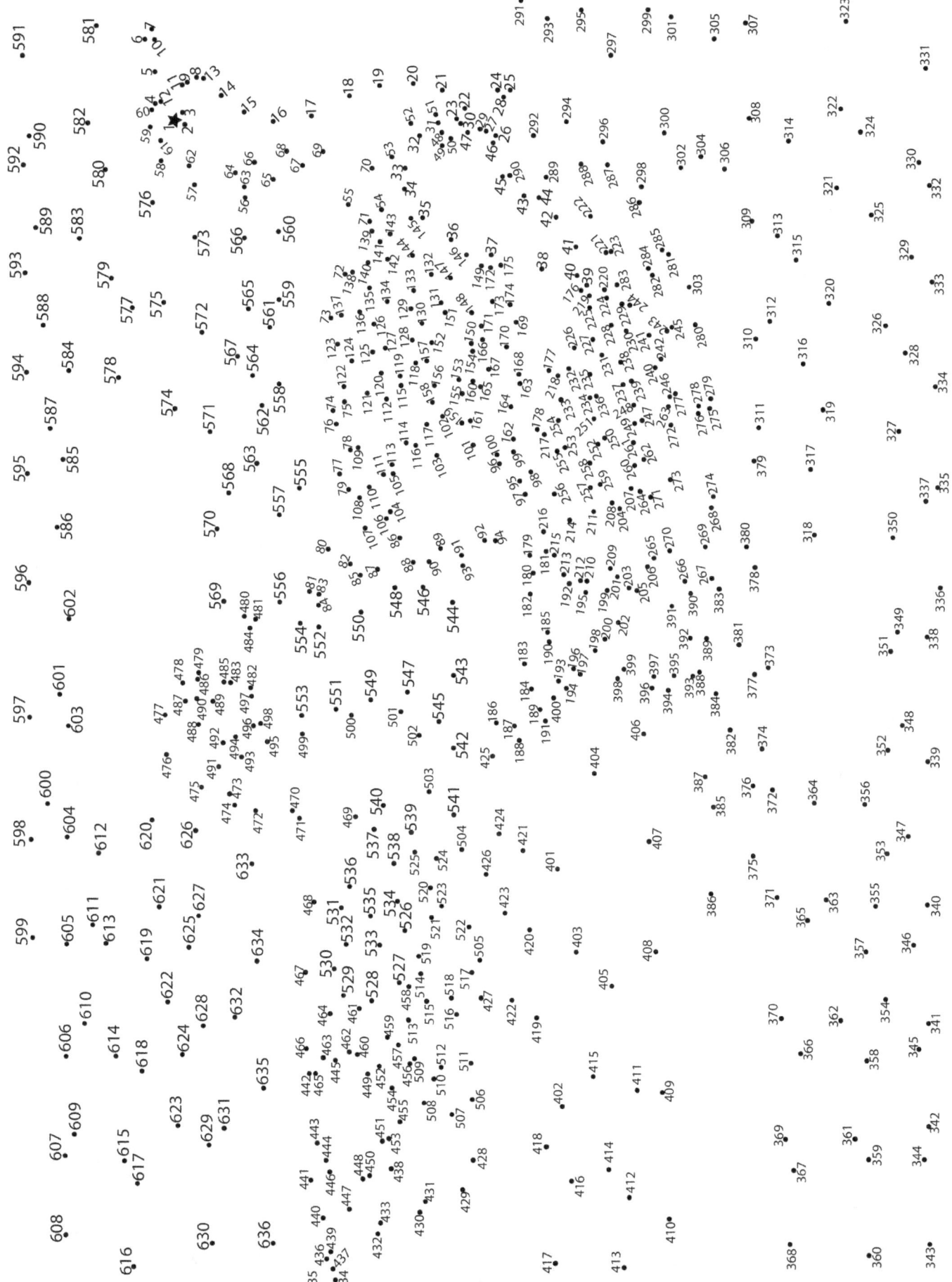

Page 11

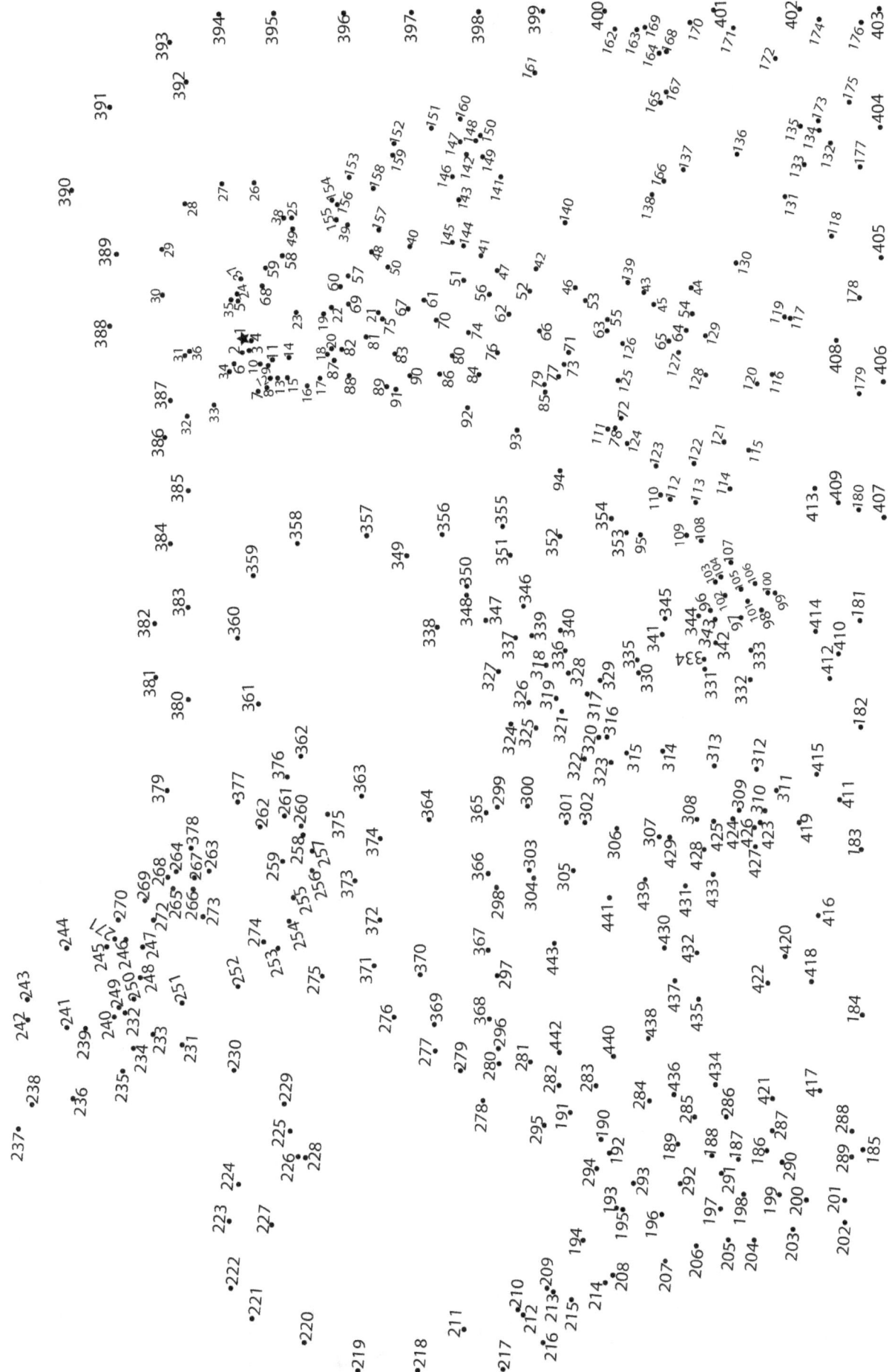

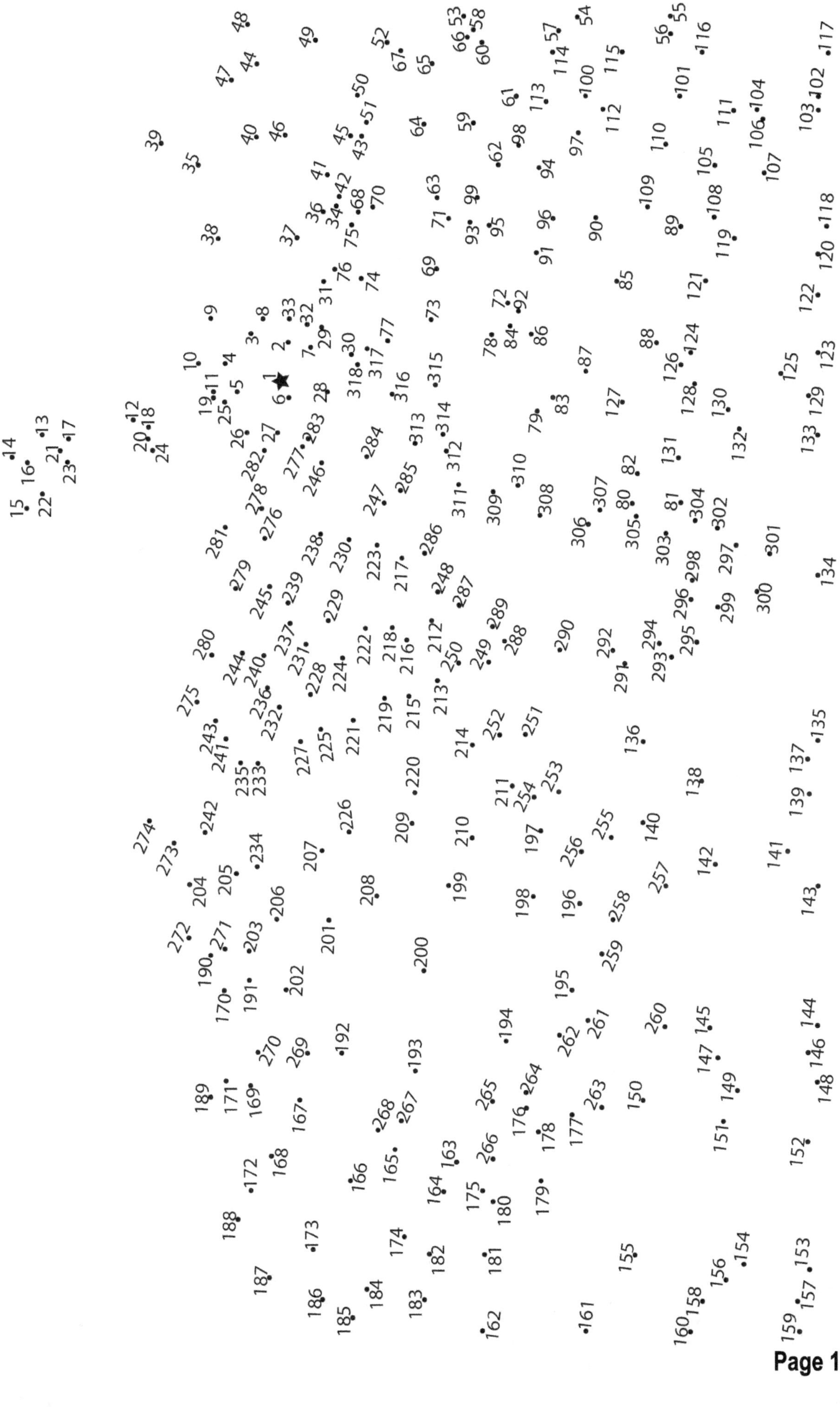

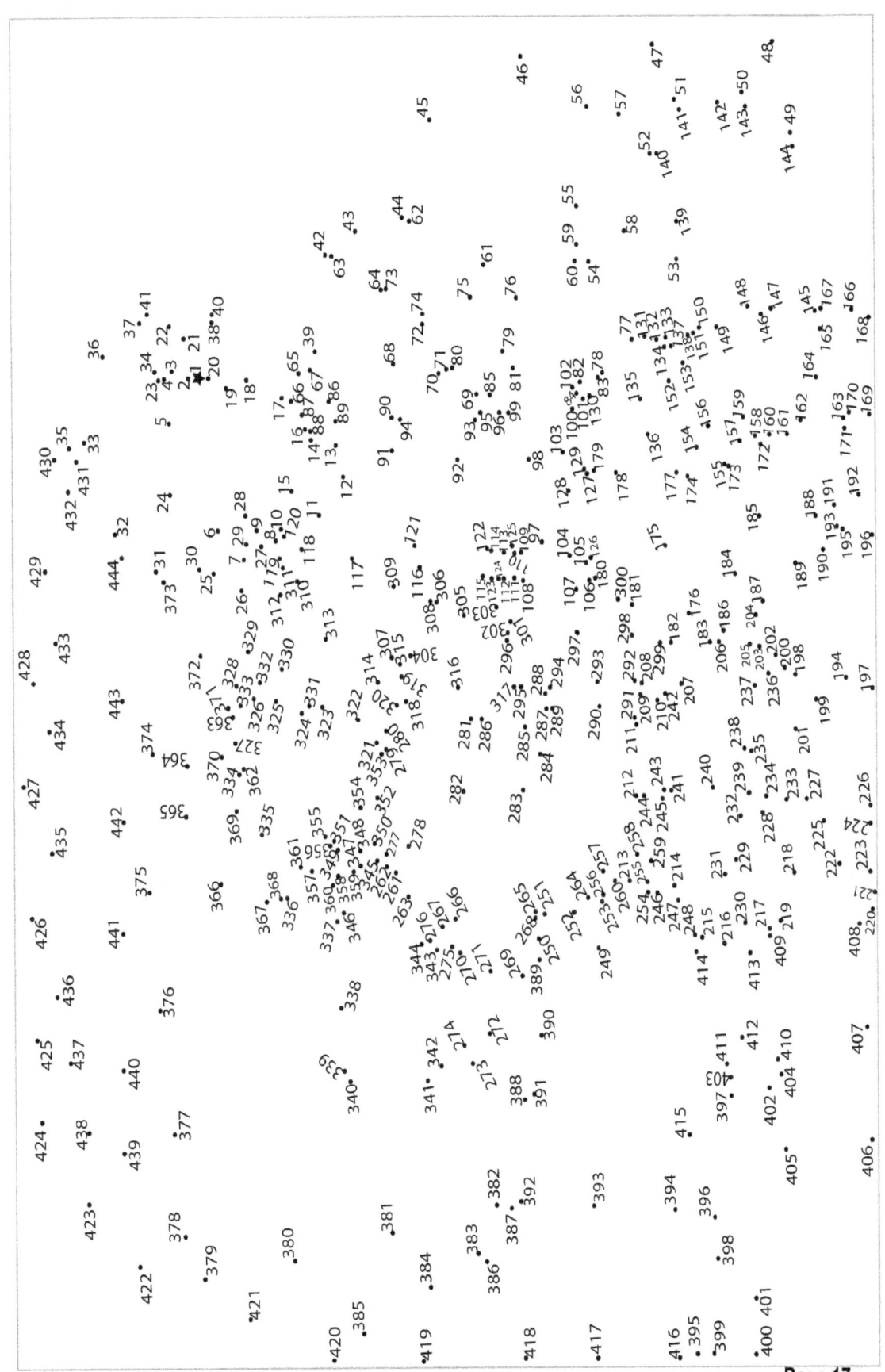

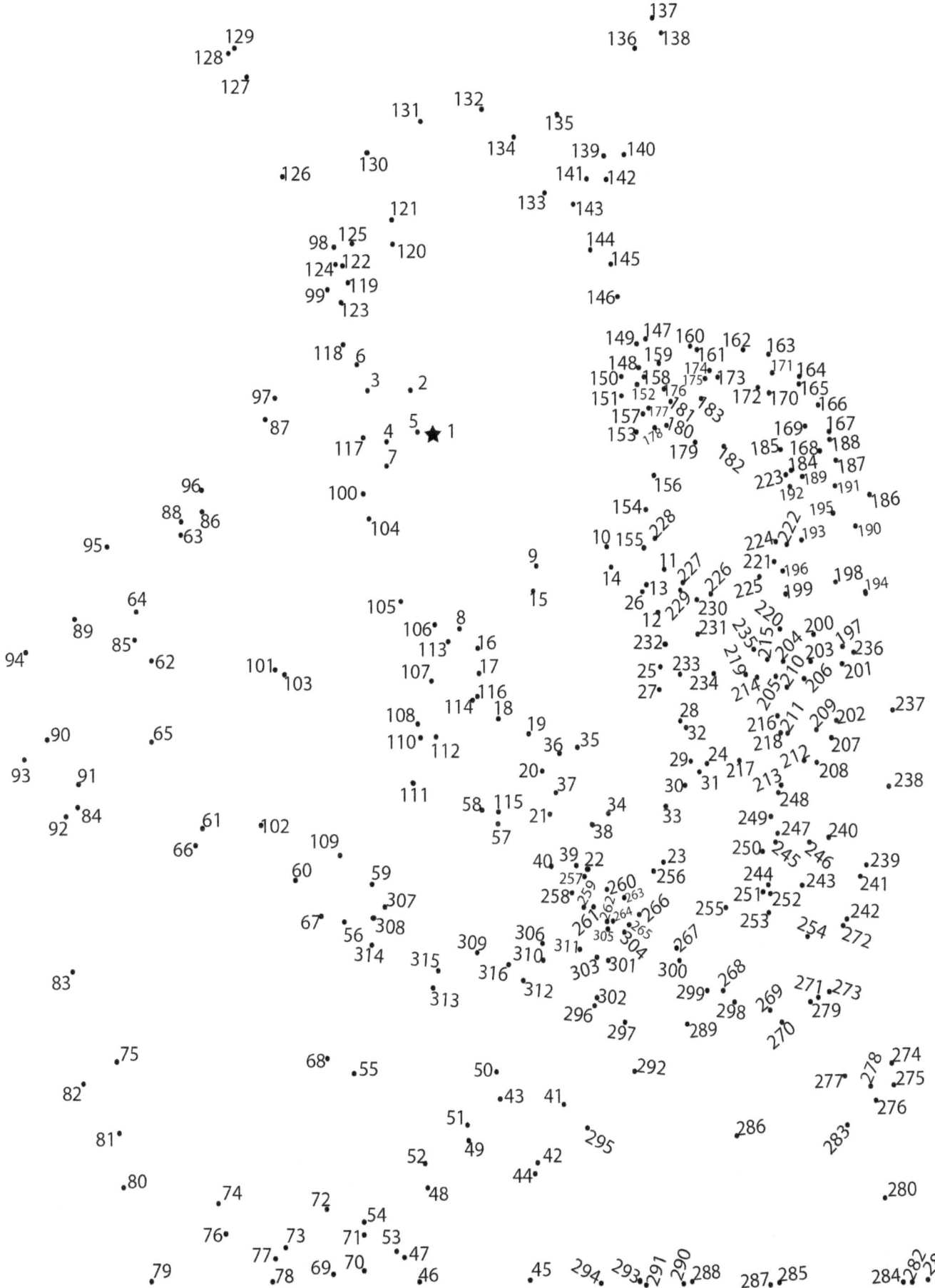

Enjoy bonus images from some of our other fun dot-to-dot books

Find all of our books on Amazon

Antique Cars and Vintage Cars
Large Print Dot-to-Dot Book for Adults

Page 22

Unicorn, Dragons, and Magical Creatures
Dot-to-Dot Puzzles from 462 to 956 Dots

Mythical Mermaid
Dot-to-Dot Book for Adults
Puzzles from 150 to 750 Dots

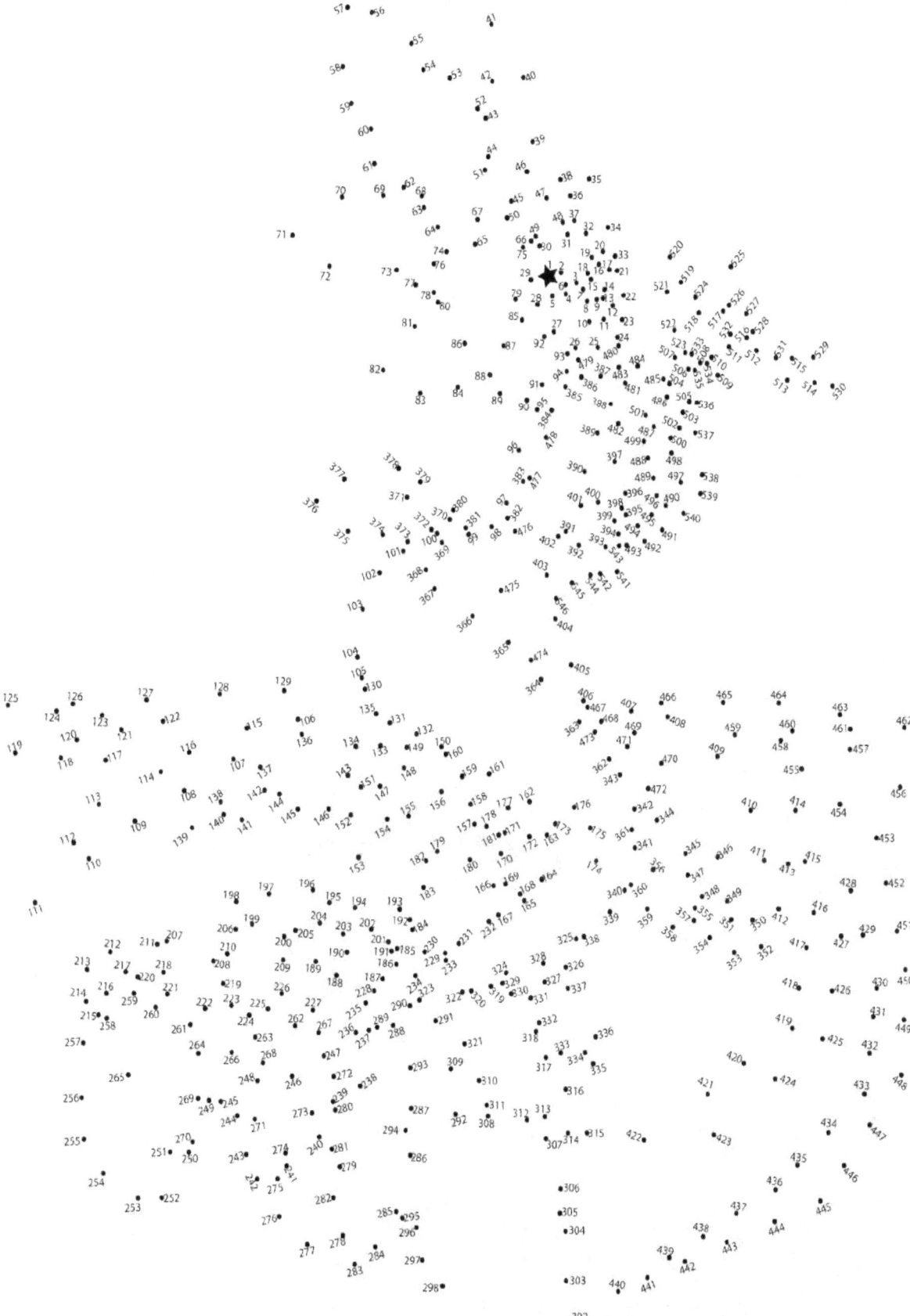

Page 24

Easy to Read Large Print Dot to Dot
Beautiful Landscapes
Puzzles from 150 to 760 Dots

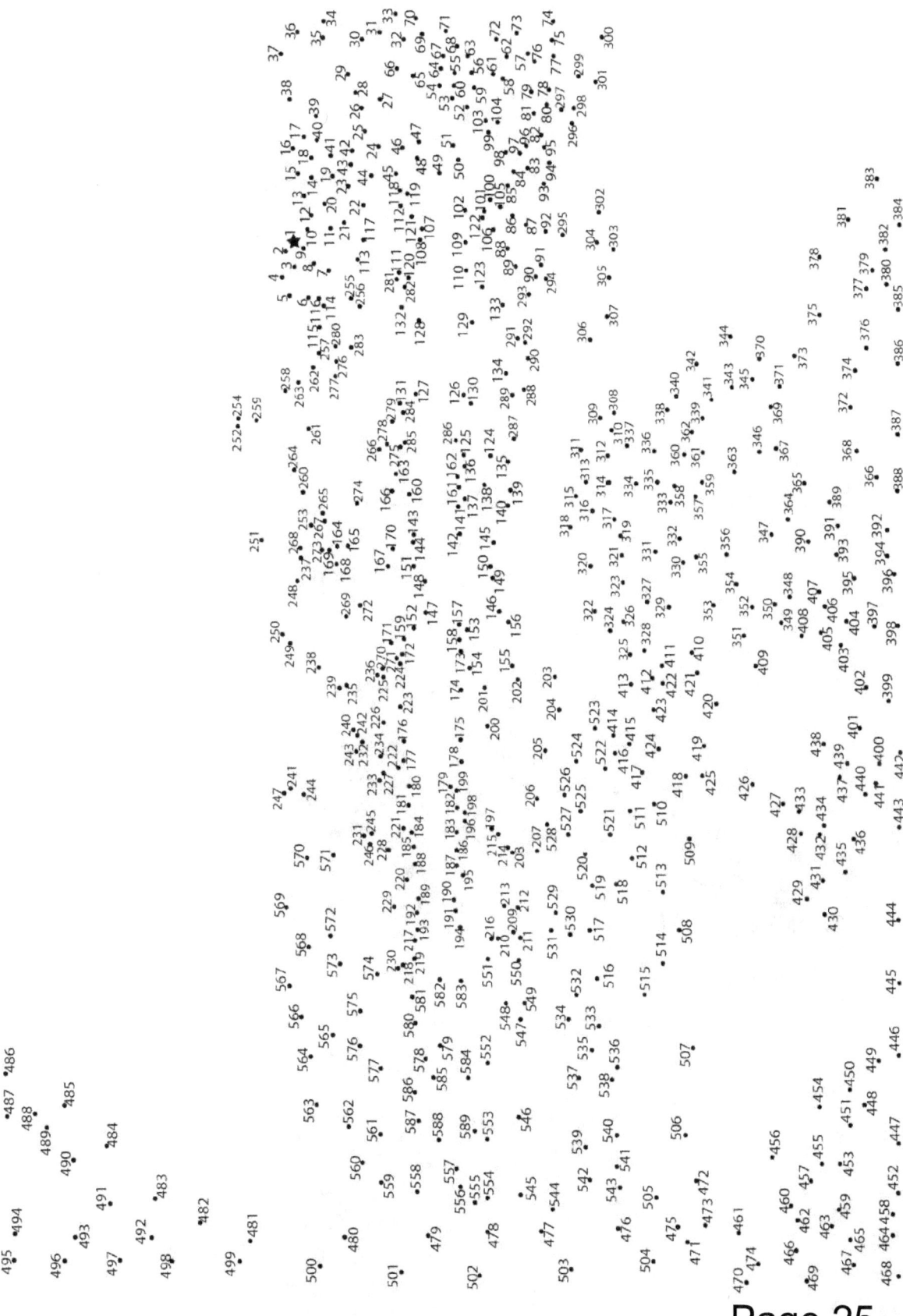

Page 25

Famous Faces
Big Book of Extreme Dot to Dot

ANSWER KEY

**Follow along with the
page numbers from top left
to bottom right**

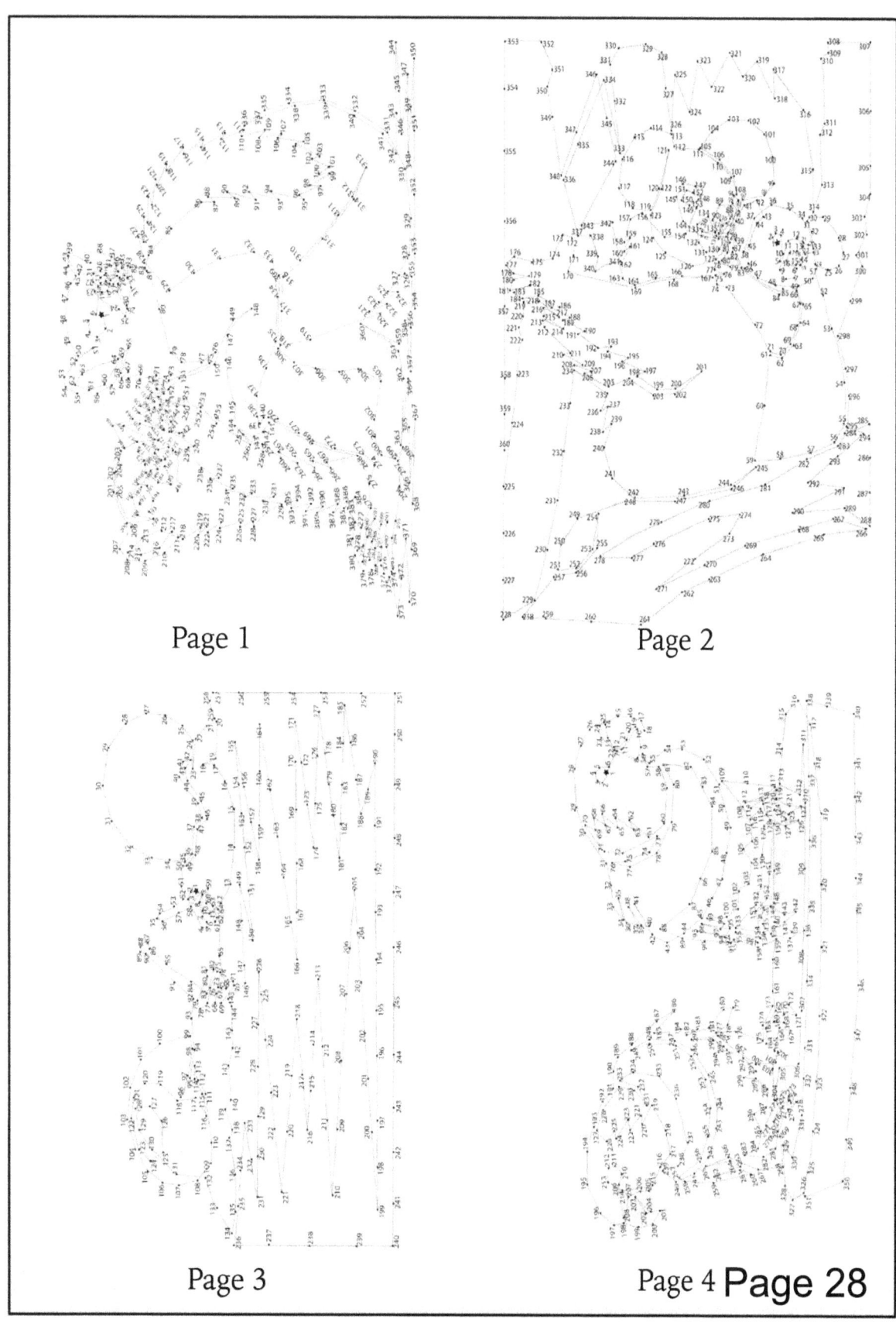

Page 1

Page 2

Page 3

Page 4 **Page 28**

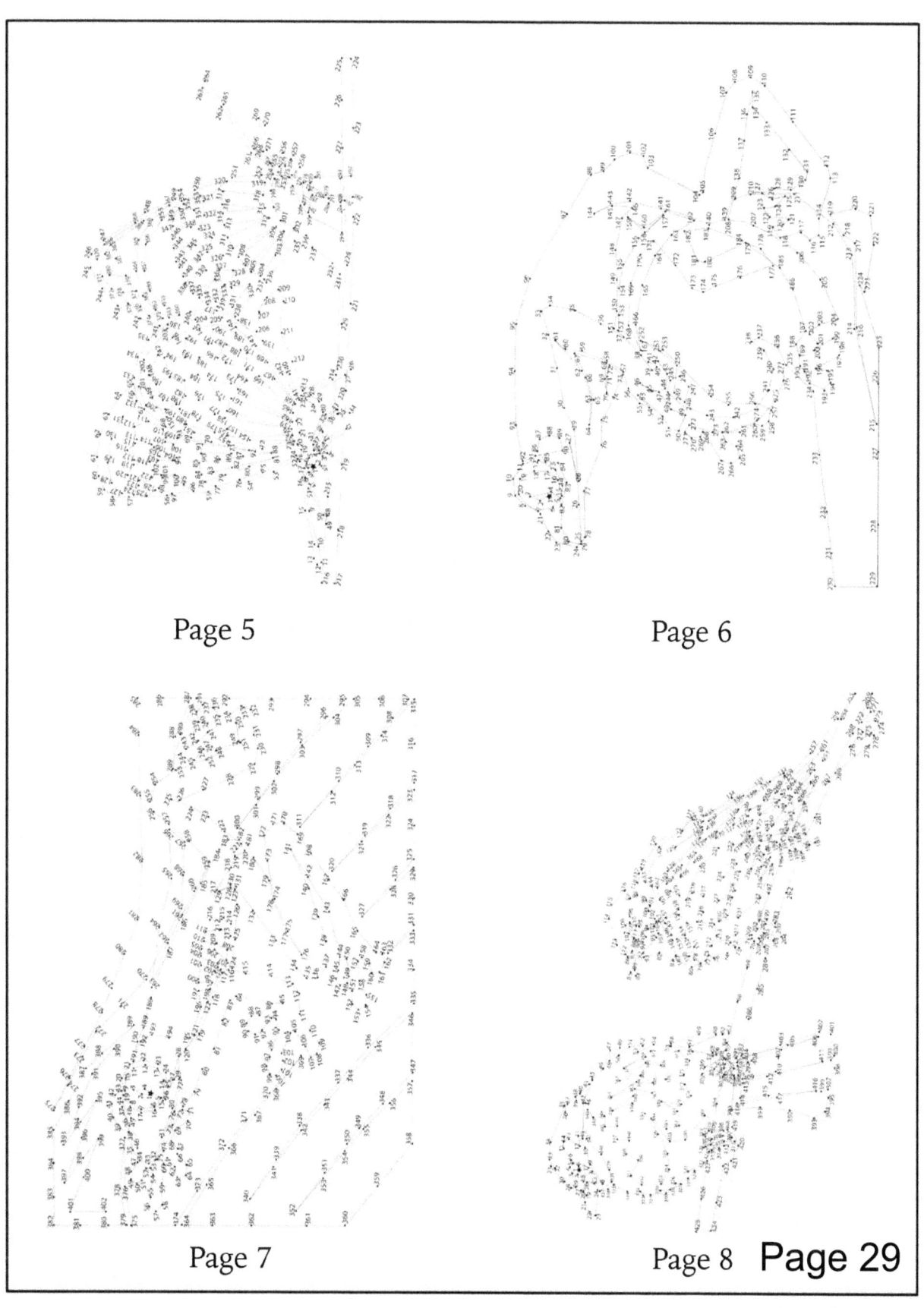

Page 5

Page 6

Page 7

Page 8   **Page 29**

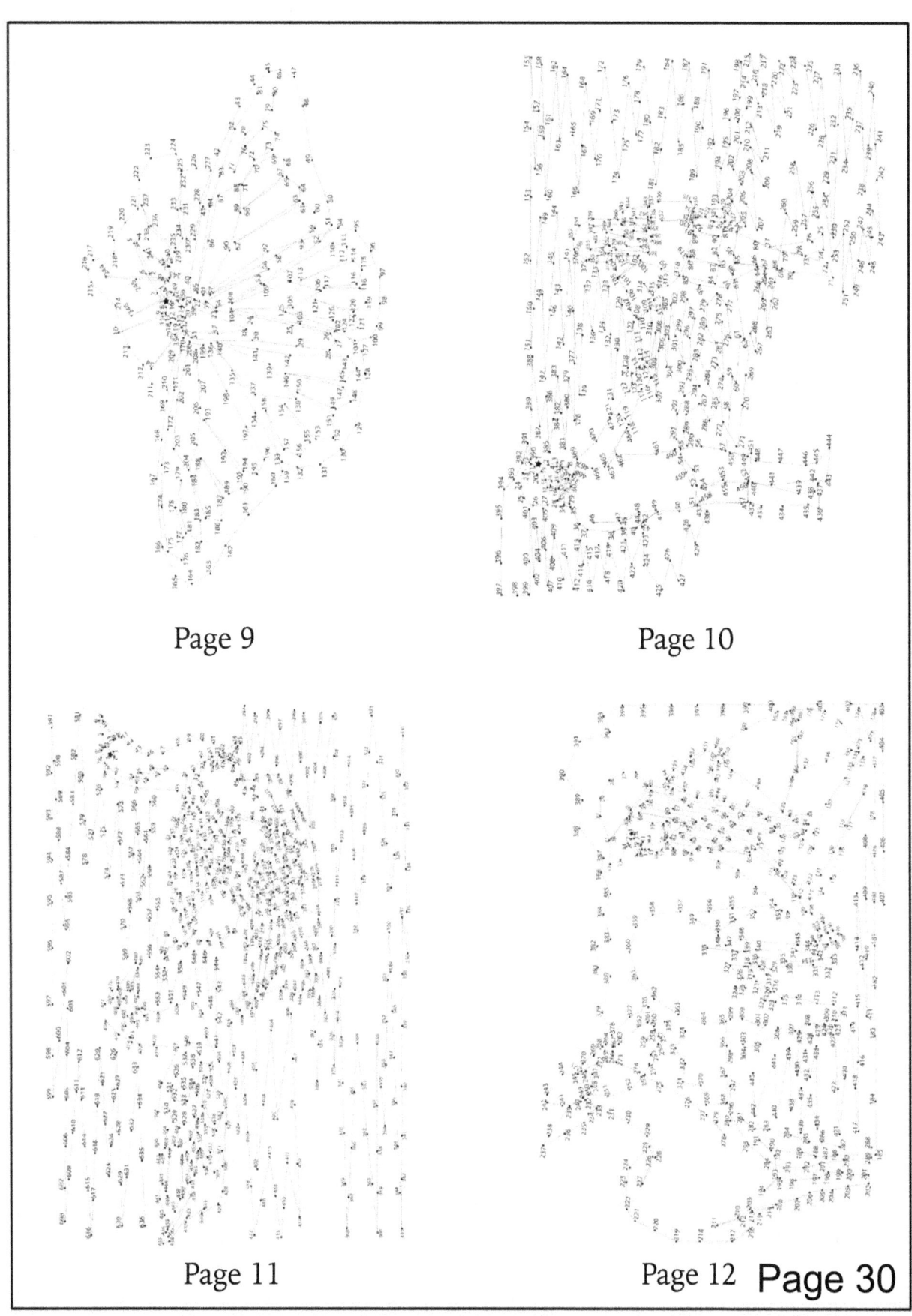

Page 9

Page 10

Page 11

Page 12 **Page 30**

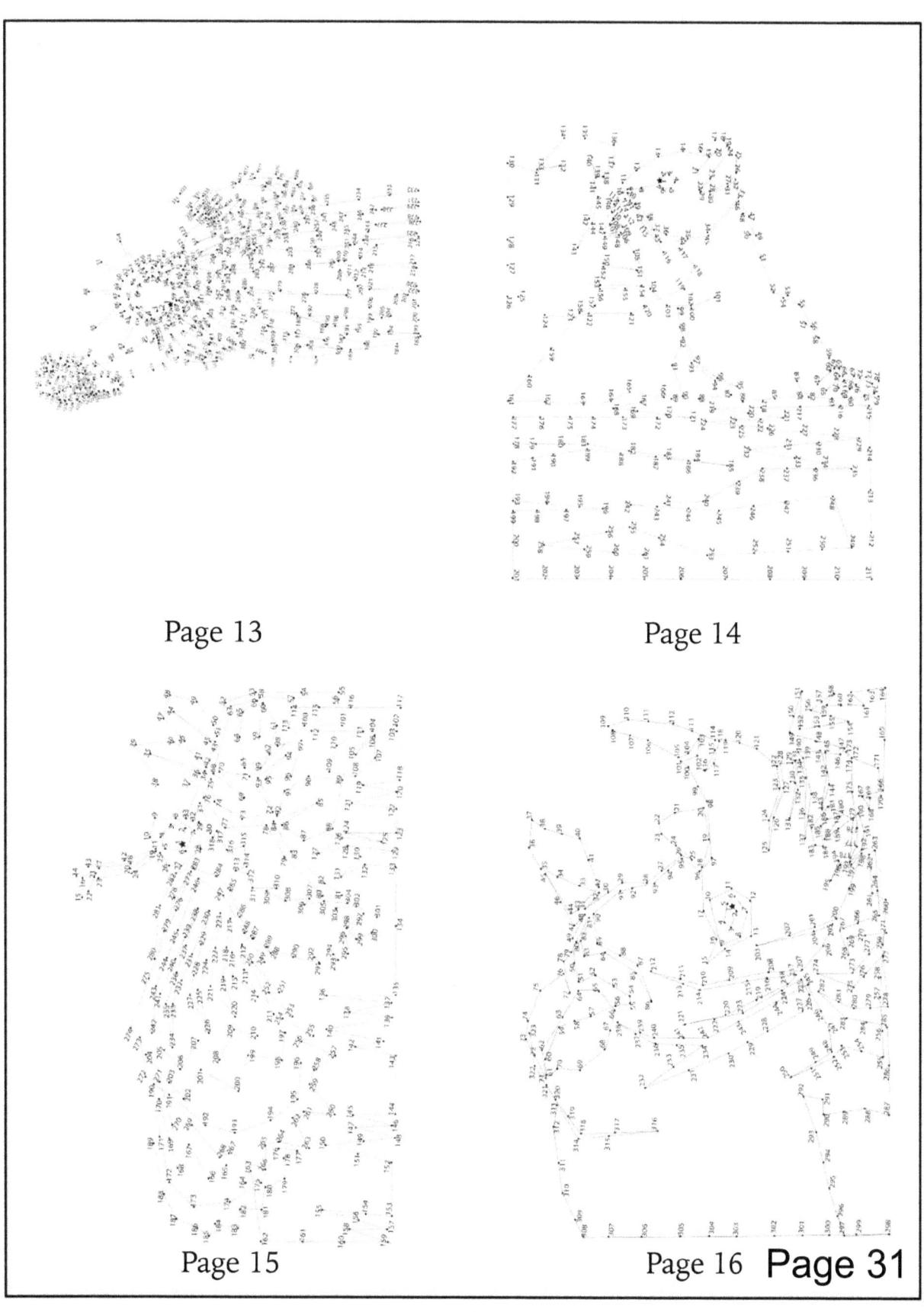

Page 13

Page 14

Page 15

Page 16   **Page 31**

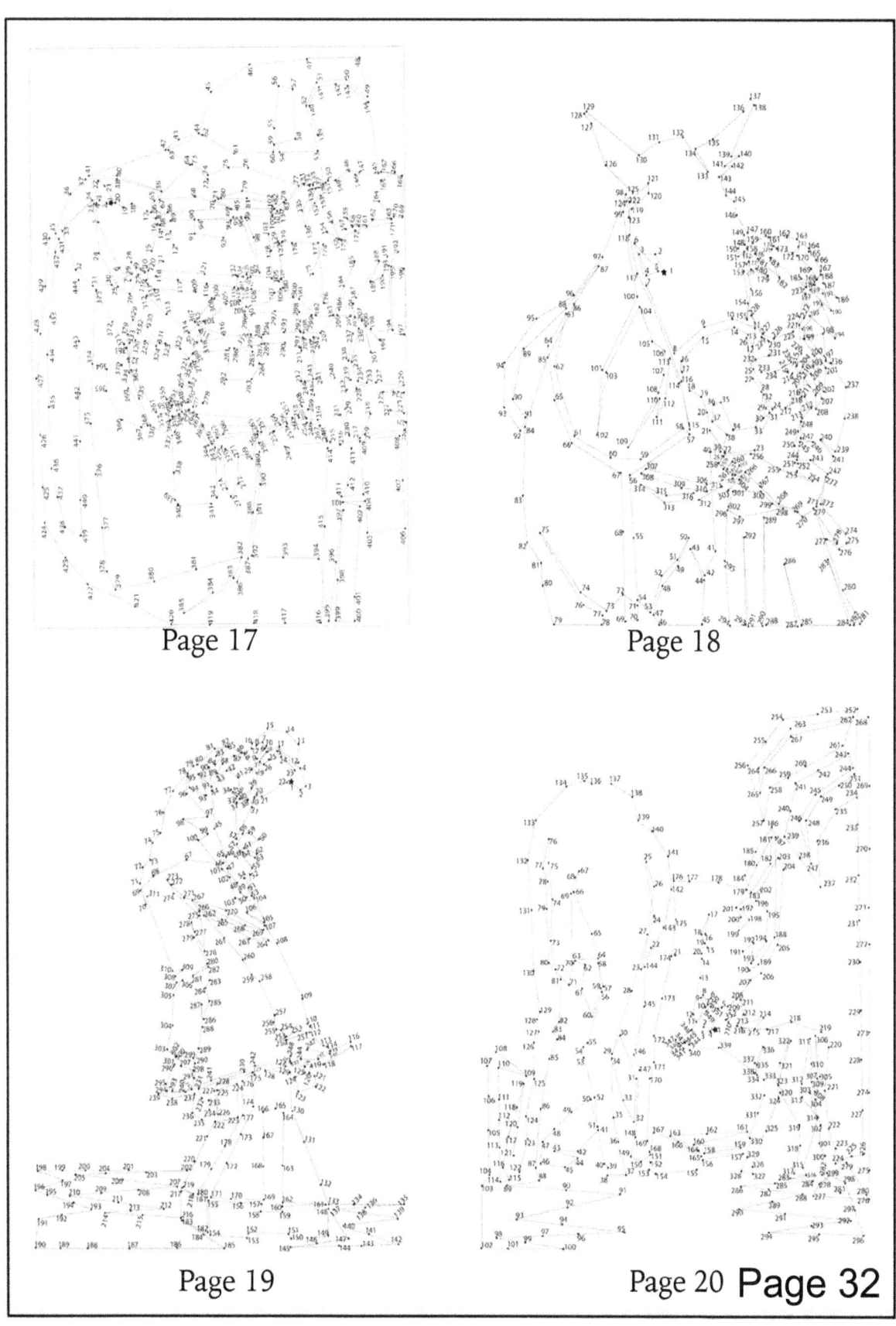

Page 17

Page 18

Page 19

Page 20 **Page 32**

www.ingramcontent.com/pod-product-compliance
Lightning Source LLC
Chambersburg PA
CBHW081750220526
45468CB00008B/2320